Yoga: 100 Key Yoga Poses and Postures Picture Book for Beginners and Advanced Practitioners
The Ultimate Guide For Total Mind and Body Fitness

By Sam Siv

Click HERE To See The Video Trailer

Legal Notice

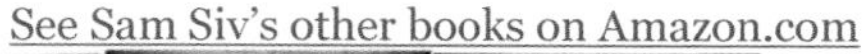

FREE YOGA TIPS!

Sign up to receive FREE Daily Yoga Tips for 32 days.

HERE'S ONE MORE THANK YOU GIFT.

Grab Your FREE BONUS Copy Of Our Special Report,
"Manifest Your Dreams!"
Sign up here and we'll send you this FREE eBook Report.

- **Discover the power of positive thinking and how you can make your dreams really come true!**
- **The simple system for utilizing your internal power system to manifest your dreams!**
- **The law of attraction is revealed! Find out how you can seize the power that lies within!**
- **How to immediately eliminate negative energy and use affirmations to turn dreams into reality!**
- **And Much, Much More!**

Table of Contents

Pose: Parivrtta Parsvakonasana Or Revolved Side Angle

INSTRUCTIONS:

To perform this pose start by standing. Stretch one leg back and Rotate the back foot in more than you do for most other standing poses, about 45 to 60 degrees. Reach your opposite arm horizontal to the ground. Place the other palm on the ground.

Pose: Supta Padangusthasana Or Reclining Hand-To-Big-Toe

INSTRUCTIONS:

To perform this pose lie on your back. Then raise your head. Stretch your arms in front of you and grab one leg. Exhale and bend the raised leg thigh into your torso.

Pose: Sarvangasana Or Shoulder Stand

INSTRUCTIONS:

To perform this pose lie on your back. Then inhale and bring your legs up until they are horizontal. Lay your arms on the ground with your palms facing down. Point your toes. Make sure your back is not touching the floor.

Pose: Ardha Shalabasana Or Half Locust Pose

INSTRUCTIONS:

To perform this pose lay down on plane surface by kipping hands under thighs. Place your chin on the ground. Look straight ahead. Keep your hands under your thighs

Pose: Prasarita Padotasana Or Wide Leddeg Forward Bend C

INSTRUCTIONS:

To perform this pose stand with your feet more than shoulder width apart. Then exhale and stretch out and parallel your arms. Interlace your fingers behind the back, or grab the opposite elbows. Then inhale and stretch up and slightly back.

Pose: Adho Mukha Savanasana Or Downward Facing Dog

INSTRUCTIONS:

To perform m this pose strand with your feet shoulder width apart. Then while stretching and bending at the waist forward. Lift your heels off of the ground when your palms touch the ground. Press your hands into the ground. Widen through the shoulder blades. Keep the neck lengthened by touching the ears to the inner arms.

Pose: Namaskar Or Salutation

INSTRUCTIONS:

To perform this pose stand straight. Keep your feet together. Bring the palms together in the prayer position. Look straight ahead and breathe slowly.

Pose: Triyak Bhujangasana Or Twisting Cobra

INSTRUCTIONS:

To perform this pose raise the upper body with the help of the arms. Then while exhaling slowly turn the head and upper body to the right and look over the right shoulder towards the left heel. Then repeat this pose with the other side.

Pose: Adho Mukha Savanasana 2 Or Downward Facing Dog 2

INSTRUCTIONS:

To perform this pose, press your hands into the ground. Arch your back. Then widen through the shoulder blades. Keep your neck lengthened by touching your ears to your inner arms. Lift your heel off of the floor and balance on fingers and toes.

Pose: Utthita Trikonasana Or Extended Triangle Pose

INSTRUCTIONS:

To perform this pose stand with your legs more than shoulder width apart. Turn one foot sideways. Take the arm on the same side as the sideways foot and touch the floor. Bend at the hip until your arms and horizontal to the floor. Repeat on the other side.

Pose: Eka Pada Karnapeedanasan Or Leg Fold Plough Pose

INSTRUCTIONS:

To perform this pose exhale while lying on your back. Then lift your hips and abdomen off the ground. Bring your legs over the head. Touch one foot to the floor and the bend the other leg back so that it is horizontal to the floor. Repeat this pose with the other leg.

Pose: Prasarita Padotasana A Or Wide Legged Forward Bend A

INSTRUCTIONS:

To perform this pose, press your hands into the ground. Then widen through the shoulder blades. Keep the neck lengthened by touching the ears to the inner arms. Remember to keep your legs shoulder width apart.

Pose: Saral Hasta Bhujangasana Or Straight Arm (Sky Face) Cobra Pose

INSTRUCTIONS:

To perform this pose, lie with on your stomach. Place your palms on the floor beside you. Inhale and straighten your arms while pushing your torso so that it is horizontal to the floor. Tilt your head back and look up at the ceiling.

Pose: Ardha Matsyendrasana Or Half Spinal Twist Pose

INSTRUCTIONS:

To perform this pose, sit on the floor. Have one leg straight in front of you and the other bent. Take the foot of the bent leg and place it over your other leg. Turn your torso away from the bent leg and place your other hand on the ground behind your torso. Look over the opposite shoulder from the leg that is bent. Then switch sides.

Pose: Ananstasana Or Side Reclining Leg Lift Pose

INSTRUCTIONS:

To perform this pose, lie on your side. Use the arm closest to the floor to support your head. Then lift your upper leg towards your torso. Grab your leg at the ankle.

Pose: Salamba Sarvangasana Or Supported Shoulder Stand

INSTRUCTIONS:

To perform this pose, lie on your back with your arms beside you. Then inhale and lift your legs, buttocks and back off or the floor. Point your toes and keep your legs straight. Use your arms to support your back. Place your upper arms on the floor and hold your hand to your back.

Pose: Urdhva Mukha Paschimottanasana Or Upward Facing Intense Posterior Stretch

INSTRUCTIONS:

To perform this pose, lie on the back. Then bring your legs straight over the head. Now grab the outside of your feet. Your legs should be almost parallel to the ground. Your lower back should be off of the floor.

Pose: Baddha Konasana Or Cobbler Pose Sitting

INSTRUCTIONS:

To perform this pose, sit with your knees bent. Place the soles of your feet together. Stretch your arms behind you keeping them straight. Place your palms on the floor. Sit up straight and look forward.

Pose: Bhujangasana Or Cobra Pose

INSTRUCTIONS:

To perform this pose, lie on your stomach. Place your palms on the ground. Start with your elbows bent then slowly stretch your arms upwards until they are straight. Hold up your body do that your torso is up right and your legs are parallel to the ground.

Pose: Halasana Or Plow Pose

INSTRUCTIONS:

To perform this pose, lie on your back. Place your arms on your sides with your palms touching the floor. Exhale and push your legs up and over your head. Stretch your legs behind your head until your toes can touch the floor. Your back should not be touching the floor.

Pose: Triyak Bhujangasana Or Twisting Cobra

INSTRUCTIONS:

To perform this pose, lie on your stomach. Raise the upper body with the help of the arms. Exhaling slowly turn the head and upper body to the right and look over the right shoulder towards the left heel.

Pose: Purna Titli Asana Or Balancing Butterfly Pose

INSTRUCTIONS:

To perform this pose, sit on the floor with your knees bent. Place the bottom of your feet together. Then place your arms behind you. Shift your weight on to your hands. Lift your behind off the floor.

Pose: Shalabhasana Or Locust Pose

INSTRUCTIONS:

To perform this pose, start by lying on your stomach with your arms by your side. Then slowly shift your weight from your legs to your arms. Raise your legs up. It helps to bend your knees slightly.

Pose: Prasarita Padotasana A Or Wide Legged Forward Bend Salutation

INSTRUCTIONS:

To perform this pose, stand with your feet shoulder width apart. Then bend your knees slightly and bend forward. As your do this lift your heels off of the ground. Lean forward until your head touches the ground and your arms are straight in front of you. Make sure your palms are touching the floor.

Pose: Paschimottanasana Or Seated Forward Bend

INSTRUCTIONS:

To perform this pose, sit on the floor with your legs in front of you. Then slightly bend your knees. Lean your torso forward and stretch out your arms. Hold your feet with your hand and keep your head down.

Pose: Sirsasana Or Headstand

INSTRUCTIONS:

To perform this pose, start by kneeling with your palms on the floor. Then lean forward and place the top of your head on the floor. Shift your weight from your knees to your hand and head. Your legs should now be up in the air. Point your toes and bend one knee.

Pose: Paschimottanasana Or Seated Knee Bent Toe Touches

INSTRUCTIONS:

To perform this pose, sit with your legs in front of you. Then slightly bend your knees. Lean forward stretching your arms and wrapping your hand around your feet. Keep your neck straight and look forward. Make sure your arms are straight.

Pose: Anantasana Or Sleeping Vishnu Pose

INSTRUCTIONS:

To perform this pose, lie on your side with your lower arm bent and supporting your head and neck. Your upper arm should be in front of you palm down on the ground. Point your toes and slowly lift your higher leg while bending your knee. You knee should reach your shoulder.

Pose: Ustrasana Or Camel Pose

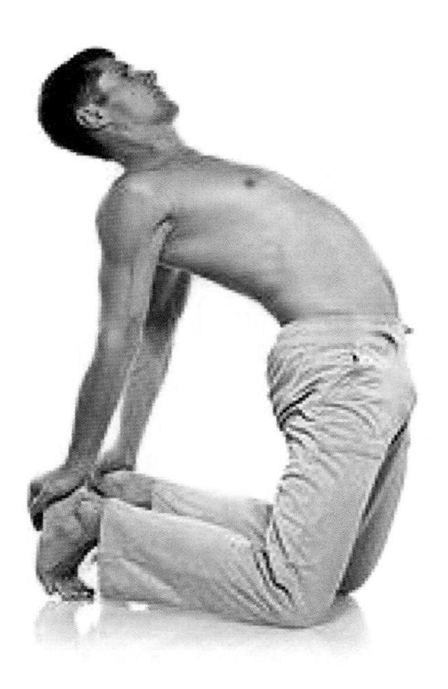

INSTRUCTIONS:

To perform this pose, kneel with both knees on the floor. Then arch your back. While doing so reach your arms behind you. Rest your hands on your knees. Keep your head back.

Pose: Pawanmuktasana Or Wind Relieving Pose In Yoga

INSTRUCTIONS:

To perform this pose, lie on your back with your toes pointed. Then slowly bring one knee up to your chest. Use your arms to hold your leg at the calf. Hold the pose and then switch legs.

Pose: Paschimottanasana Or Seated Forward Bend

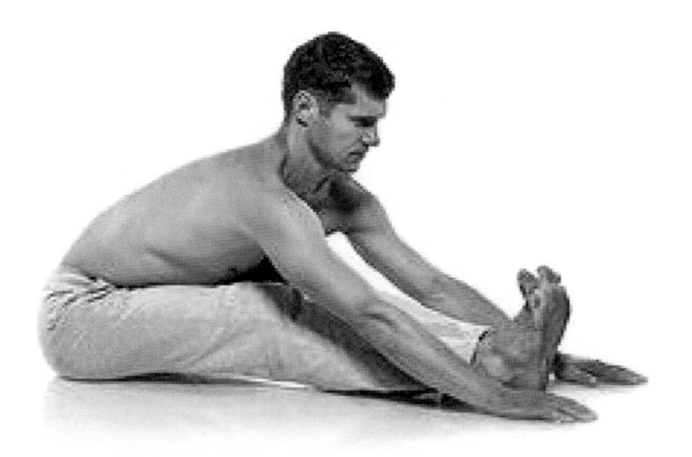

INSTRUCTIONS:

To perform this pose, sit on the floor with your toes point upwards. Slowly lean forward. Keep your arms in front of you. Lean forward until your torso is touching your legs. Stretch your arms forward until your palms are touching the floor and your arms are straight.

Pose: Eka Pada Rajakapotasana Or One Legged King Pigeon Pose

INSTRUCTIONS:

To perform this pose, get down on one knee. Then using your arms grab the foot of the leg that has a knee on the ground. Bend that knee until your elbows are making a 90 degree angle. Keep your head up looking straight.

Pose: Salamba Bhujangasana Or Sphinx Pose

INSTRUCTIONS:

To perform this pose, lie on your stomach. Place your arms in front of you with your palms flat on the ground. Then slowly lift your torso off the ground until your arms are completely straight. Remember to keep looking towards the ground.

Pose: Dhanurasana Or Bow Pose

INSTRUCTIONS:

To perform this pose, lie on your stomach. Then slowly bend your knees. Lift your knees and calves off the floor until you are able to reach them with your arms. Straighten your arms and keep your neck straight.

Pose: Parivrtta Tikona Or Twisted Triangle Pose Yoga

INSTRUCTIONS:

To perform this pose, spread your feet far apart. Then forward bend at the waist. As you bend straighten your arms to your sides. Reach the opposite arm to the opposite leg. Repeat for each side.

Pose: Bhujangasana Or Cobra Pose

INSTRUCTIONS:

To perform this pose, lie on your stomach. Then place your hands beside you with bent elbows. Slowly push your body up until your arms are straight. Continue to look at the ground.

Pose: Prasarita Padottanasana D Or Wide Legged Forward Bend D

INSTRUCTIONS:

To perform this pose, keep your feet flat on the ground and spread your legs wide apart. Slowly bend downwards at the waist until your hands can wrap around your ankles. Grab your ankles from behind.

Pose: Dandayamana Dhanurasana Or Standing Bow Pulling Pose

INSTRUCTIONS:

To perform this pose, stand on one leg. Then with the opposite arm grab the opposite leg at the ankle. Stretch the leg behind you keeping your knee bent. Lean your torso forward. Raise your opposite arm straight beside your head.

Pose: Chandrasana Or Crescent Moon Pose

INSTRUCTIONS:

To perform this pose, lunge forward with one leg and bend at the knee. With the opposite leg, stretch it as far back as it can go and place your knee to your foot on the ground. Then raise your arms straight above your head with your palms facing each other.

Pose: Vakra Hasta Bhujangasana Or Curved Hands Cobra Pose

INSTRUCTIONS:

To perform this pose, start by lying on your stomach. Then using your hand push your torso up. Hold one hand in front of the other. Keep your head facing downward. You can switch sides.

Pose: Prasarita Padottanasana B Or Wide Legged Forward Bend B

INSTRUCTIONS:

To perform this pose exhale and stretch out and parallel your arms. Interlace your fingers behind your back, or grab the opposite elbows. Then inhale and stretch up and slightly back.

Pose: Hanumanasana Or Splits Pose

INSTRUCTIONS:

To perform this pose, start by squatting down. Slowly stretch your legs in opposite directions. When you get low enough use your arms for support on the floor. Keep your torso straight. Your foot can be turned to the side.

Pose: Pose: Bhekasana Or Frog Pose

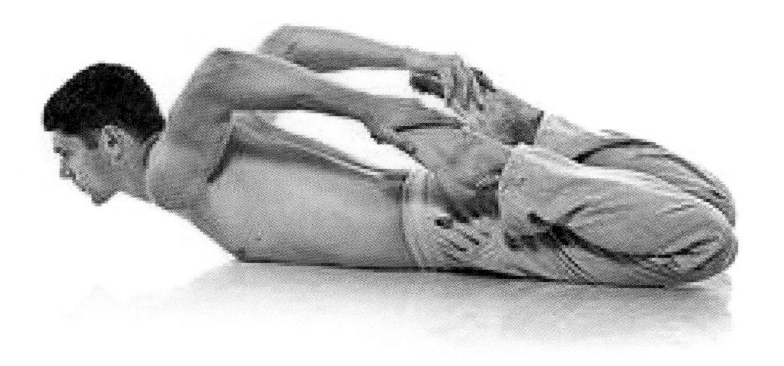

INSTRUCTIONS:

To perform this pose, start by lying on your stomach. Then bend your knees so that your feet are touching your buttock. Then bring your arms back behind you and grab your feet. Your elbows should be at a 90 degree angle and you should raise your head, neck and shoulders off the ground.

Pose: Balasana Or Child's Pose

INSTRUCTIONS:

To perform this pose, kneel on the floor on all fours. Slowly bring your hands back towards your knees. Then lower your head to the ground. Remember to keep your elbows bent.

Pose: Mukta Hasta Sirsasana Or Free Hands Head Stand

INSTRUCTIONS:

To perform this pose, star kneeling on all fours with your elbows on the ground. Then slowly shift your weight from your knees to your elbows as you lift your knees off the ground. Slowly straighten your legs. Do not try to lift your head.

Pose: Ardha Uttanasana Or Half Standing Forward Bend Pose

INSTRUCTIONS:

To perform this pose, stand with your feet more than shoulder width apart. Crouch forward like you are sitting in a chair. Lean your torso forward until it is parallel with the ground. Lift your head and place your hands on your knees. Your arms should be bent.

Pose: Vajrasana Or Diamond Pose

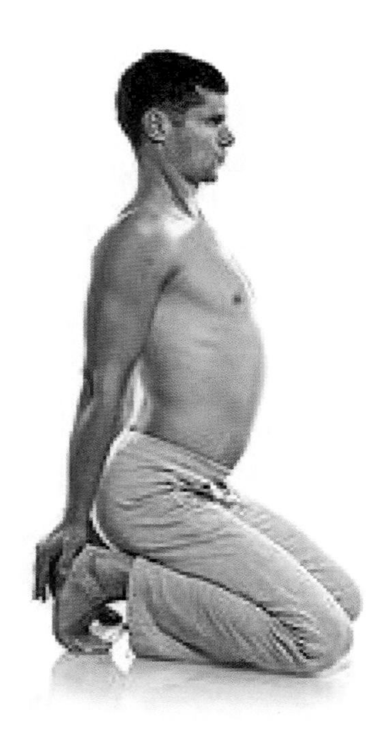

INSTRUCTIONS:

To perform this pose, start in a kneeling position. Then place your arms straight by your side and behind you. Place the palms of your hands on your heels. Keep your shoulders down and look forward.

Pose: Salabhasana Or Locust Pose Hands On Head

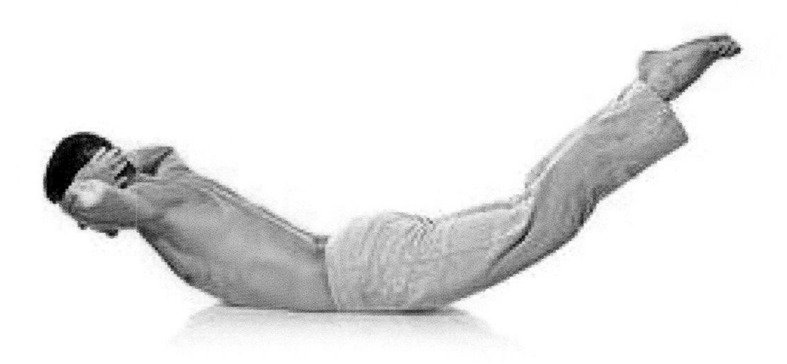

INSTRUCTIONS:

To perform this pose, start by lying on your stomach. Then lift your legs up. Slightly bend your knees. Now place your hand on the back of your head with your elbows bent and lift your upper body off of the ground.

Pose: Bhekasana 2 Or Frog Pose 2

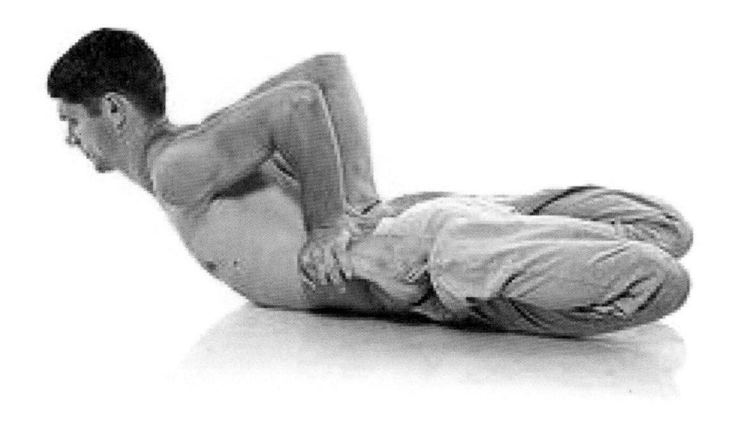

INSTRUCTIONS:

To perform this pose, lie on your stomach. Then bend your knees until they are at your sides. Bring your arms back so that they are holding your feet. Your elbows should be bent at a 90 degree angle.

Pose: Salamba Sirsasana Or Teddy Bear Headstand

INSTRUCTIONS:

To perform this pose, start on all fours. Then tuck your knees into your arms and lean forward. Bending your elbows. Your head should be touching the floor and your elbows should be bent at a 90 degree angle.

Pose: Supta Virasana Or Reclining Hero Pose

INSTRUCTIONS:

To perform this task, kneel with your buttock touching your feet. Now lean backwards. Your head should be touching the floor. Your arms should be wrapped around your head

Pose: Pavana Muktasana Or Wind Relieving Pose

INSTRUCTIONS:

To perform this pose, sit on the floor. Then bring your knees up to your chest. Wrap your arms around your legs and roll backwards. Your shoulders and legs should be on the floor, while your lower back, buttocks and legs should be up in the air

Pose: Ashva Sanchalanasana (A) Or Horse Riding A

INSTRUCTIONS:

To perform this pose, sit on the floor. Then wrap your arms around your knees. Lean back a little and lift your feet off of the floor.

Pose: Urdhva Upavistha Konasana Or Upward-Facing Open Angle Pose

INSTRUCTIONS:

To perform this task, Like on your back. Stretch your arm out to the sides. Now bring your legs up towards your torso. Do this until your hands are able to hold your feet. Do not bend your arms.

Pose: Setu Bandhasana Or Bridge Pose

INSTRUCTIONS:

To perform this pose, lie on your back with your arms at your side. Then place your feet flat on the floor and push your body up. Do this until your toes are touching the floor and your thighs are parallel to the floor.

Pose: Naukasana Or Boat Pose

INSTRUCTIONS:

To perform this pose, lie on your back. Then slowly lift your back off of the floor with your arms straight in front of you. The do the same with your legs and point your toes. Your arms and legs should be at 45 degree angles.

Pose: Ananda Balasana Or Happy Baby Pose

INSTRUCTIONS:

To perform this pose, lie on your back. Now bend your knees towards your body until your knees are touching your armpits. Use your arms to hold your ankles. This will keep your legs in place.

Pose: Laghu Vajrasana Or Little Thunderbolt

INSTRUCTIONS:

To perform this pose, kneel on the floor. Then lean backwards. Do this until your buttock is touching your feet. Then place your arms behind you. Your forearms and palms should be touching the floor.

Pose: Prasarita Padottanasana A Or Wide Legged Forward Bend A

INSTRUCTIONS:

To perform this pose, start by kneeling with your feet flat on the floor. Then place your palms on the floor. Now lift your buttock into the air and place your forehead on the floor.

Pose: Paschimottanasana Or West Stretching Bend

INSTRUCTIONS:

To perform this pose, sit on the floor with your legs together. Then bend your torso forward. Do this until your face is touching your legs then stretch your arms forward. Make sure your forearms are touching the floor.

Pose: Ardha Purvottanasana Or Table Top Pose

INSTRUCTIONS:

To perform this pose, lie on your back with your feet flat on the floor and your knees bent. Now place your palms on the floor and push your body up until your arms are extended and your torso is parallel to the ground.

Pose: Janushirasana Or Head To Knee Pose

INSTRUCTIONS:

To perform this pose, sit on the floor and bend one leg in. The other leg should be straight. Now bend your torso so that you are face down. Make sure your opposite hand is touching the floor.

Pose: Utthita Marjaryasana Or Extended Cat Pose

INSTRUCTIONS:

To perform this pose, kneel on all fours. Lift your palms off of the floor. Then lift one leg out to the side. Make sure it is straight. Turn your head to the side of the leg that is out.

Pose: Bibhaktapada Janushirasana Or Sep Leg Head To Knee

INSTRUCTIONS:

To perform this pose, start by kneeling on one knee. Then stretch the other leg in front of you so that your leg is straight and your foot is flat on the floor. Now bend your torso until it is touching that leg. Use your arms for balance.

Pose: Chaturanga Asana Or Plank Pose

INSTRUCTIONS:

To perform this pose, lie face down on the floor as if you are about to do a push up. Now push your body up with the palms of your hands and your toes. Hold this pose for 20 seconds to 1 minute. Make sure your neck is parallel to the floor.

Pose: Eka Pada Marjariasana Or Single Leg Cat Pose

INSTRUCTIONS:

To perform this pose, kneel on all fours. Then bring one leg in so that your knee is touching your chin and your toes are on the floor. Now lift your palms off the floor.

Pose: Ardha Baddha Padmasana Or Half Bound Lotus Pose

INSTRUCTIONS:

To perform this pose, sit on the floor with one leg extended in front of you. Bend the other leg so that the bottom of your foot in touching your other leg. Now place your hands behind you so that your palms are flat on the floor behind you.

Pose: Utthita Marjaryasana Or Extended Cat Pose

INSTRUCTIONS:

To perform this pose, kneel on all fours. Lift up the palms of your hand so that your finger tips are touching the floor. Not lift one leg straight up. Hold this pose and look up at the ceiling. Change sides.

Pose: Paschimottanasana B Or Seated Forward Bend Pose B

INSTRUCTIONS:

To perform this pose, sit on the floor with your legs together. Then slightly bend your knees so that they do not touch the ground. Lean your torso forward until it touches your thighs. Now stretch your arms forward and place your palms flat on the floor. Look at your feet.

Pose: Dynamic Majariasana Or Relaxing Cat Pose

INSTRUCTIONS:

To perform this pose, kneel on the floor. Then place your arms in front of you with your fingers touching the floor. Now look down at the ground. Your back should be curved forward.

Pose: Majariasana Or Cat Stretch Pose

INSTRUCTIONS:

To perform this pose, kneel on the floor. Then arch your back and place your finger tips on the floor. Look up at the ceiling and stretch your neck.

Pose: Ardha Chandrasana Or Half Moon Pose

INSTRUCTIONS:

To perform this pose, stand on one foot. Then bend your torso at the waist until your hand touches the floor. This should be the same hand as your foot that is on the floor. Now raise the opposite leg until it is higher than the rest of your body. Switch sides.

Pose: Utkatasana Or Awkward Chair Pose

INSTRUCTIONS:

To perform this pose, bend your knees slightly. Then stand on your toes. Lean your torso back and interlock our fingers. Stretch your arms forward so that they are parallel to the ground.

Pose: Garudasana Or Eagle Pose

INSTRUCTIONS:

To perform this pose stand on one foot with your knee bend. Wrap you other leg over and around the bent leg. Then bend your torso towards your knees. Now wrap your arms similar to your knees and place your palms together.

Pose: Parivrtta Trikonasana Or Revolved Triangle Pose

INSTRUCTIONS:

To perform this task, stand on one foot and lift your other leg and bend it at a 90 degree angle. Then with your opposite arm hold your knee and bend your leg towards your other leg. Use the opposite arm for balance.

Pose: Surya Namaskar B Or Sun Salutation B

INSTRUCTIONS:

To perform this pose, stand with your feet together. Then bend your knees slightly. Then stretch your arms straight up into the air. Now place your hands together so that your palms are touching.

Pose: Hastapadasana Or Standing Forward Bend

INSTRUCTIONS:

To perform this pose, stand straight with your feet shoulder length apart. Then bend at the waist until your torso is parallel to the ground. Then stretch your arms in front of you with your palms touching each other.

Pose: Janushirasana Or Seated Head To Knee Pose

INSTRUCTIONS:

To perform this pose, sit on the floor with your legs in the shape of a "V". Then bend one leg towards your other leg. Put your arms behind you for support and bend your torso towards your straight leg. Make sure your chin is touching your leg. Then switch sides.

Pose: Uttana Shishosana Or Extended Puppy Dog Pose

INSTRUCTIONS:

To perform this pose, kneel on the floor with your knees shoulder with apart. Then bend your torso towards the ground. Touch the tips of your fingers to the floor and look forward.

Pose: Namaskara Or Salutation

INSTRUCTIONS:

To perform this pose you should be standing straight with your feet together. Then bend your elbows and bring your hand together. Touch your palms together in a prayer position.

Pose: Utthita Hasta Padangusthasana Extended Hand To Big Toe Pose

INSTRUCTIONS:

To perform this pose, stand on one foot. Grab the opposite foot with the same side hand and stretch it up until your toes are pointing to the ceiling. Your arms should be straight out from your body. Your opposite arm should be straight out.

Pose: Salamba Kapotasana Or Supported Pigeon Pose

INSTRUCTIONS:

To perform this pose, put on leg behind you with your toes touching the floor. Then bend your other leg so that it is in front of you. Place your hands on the floor for support. Look at the floor and hold the pose.

Pose: Eka Pada Rajakapotasana Or Standing One Legged King Pigeon Pose

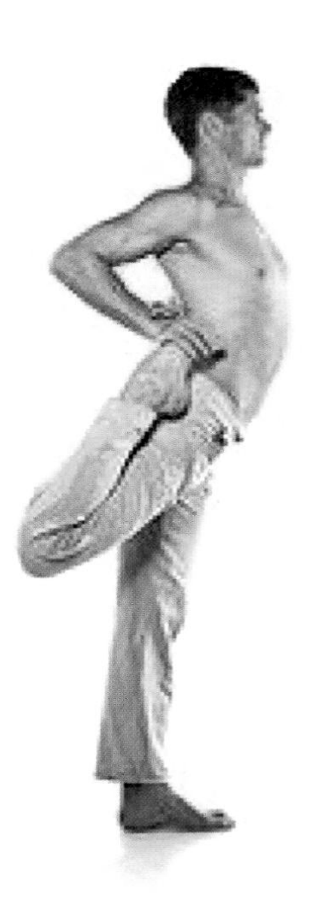

INSTRUCTIONS:

To perform this pose, stand on one leg. Then bend your other leg back towards your buttock until your foot touches it. Now hold your foot in place with your hand. Bending your elbow back behind your body.

Pose: Utkata Konasana Or Goddess Pose

INSTRUCTIONS:

To perform this pose, open your legs and bend then to a 90 degree angle. Then bend your elbows and interlock your fingers behind your head. Next bend your torso on one side so that your elbow points to the ground. Then switch sides.

Pose: Parsva Tadasana Or Twisted Leg Sidebending Mountain Pose

INSTRUCTIONS:

To perform this pose, stand on one foot and place the other leg in front of you. Touch the top of your toes to the floor and stretch your arms over your head. Interlock your fingers and bend your torso to the right and hold. Then switch legs and bend to the other side.

Pose: Virabhadrasana 1 Or Warrior 1

INSTRUCTIONS:

To perform this pose, take a big step with one foot and then bend your knee. Take the other leg and stretch it behind you. Lift your heel off of the ground. Then place your arms in prayer pose. Stretch your arms and torso back.

Pose: Garudasana Or Downward Eagle Pose

INSTRUCTIONS:

To perform this pose, stand on one foot and bend that knee. Then take the other leg and cross it over you it over the bent knee. Lean your torso forward and hold out your arms with your fingers pointing to the ground. Switch sides.

Pose: Shirangushthasana Or Sideways Bend Towards The Toes

INSTRUCTIONS:

To perform this pose, bend your left leg forward and stretch your right leg back. Do not put your right heel down. Bend your torso towards the ground and place your lower arms, palms and elbows on the ground. Make sure that your forehead touches your hands.

Pose: Utkatasana Or Awkward Chair Pose

INSTRUCTIONS:

To perform this task, bend your knees slightly as if you were trying to sit in a chair. Then lift your heels and curve your back while stretching your arms forward. Make sure your palms are facing away from your face. Then interlock your fingers.

Pose: Hasta Utthanasana Or Flying Eagle Pose

INSTRUCTIONS:

To perform this pose, stand with your legs in an inverted "V" with your feet pointed outwards. Raise your arms and keep your back straight. Bend your elbows and hold your hand behind your head. Look straight ahead.

Pose: Parivrrtta Trikonasana Or Twisting Triangle Pose

INSTRUCTIONS:

To perform this pose, stretch your left foot forward and keep it flat on the ground. Stretch your right foot back and balance on your toes. Then turn your torso and bend forward so that your extended right arm extends past your knee and the tips of your fingers can touch the ground. Turn your head to the ceiling and extend your other arm straight up so that your fingers are pointing to the ceiling.

Pose: Padahastasana Or Hand At Foot Pose

INSTRUCTIONS:

To perform this pose, stand with your feet together. Then bend your torso until your chest and face are touching your knees. Finally wrap your arms around your legs and hold on to your ankles to help you stay in the pose.

Pose: Dandayamana Bibhaktapada Janushirasana Or Sep Leg

Head To Knee Pose

INSTRUCTIONS:

To perform this pose, stand with one foot in front of the other so that your legs make an inverted "v" shape. Point your toes to the same side as your forward facing leg. Then bend your torso toward your forward facing leg and raise your arms straight behind you. Make sure you interlace your fingers and point your index fingers to the ceiling.

Pose: Hastapadasana Or Standing Forward Bend

INSTRUCTIONS:

To perform this pose, stand is a half squat position. Then bend your torso forward until it is parallel with the ground. Next stretch your arms forward behind your head. Make sure your palms are touching in a prayer position.

Pose: Parivrtta Parsvakonasana Or Revolved Extended Side Angle Pose

INSTRUCTIONS:

To perform this pose stand in the warrior pose. Over your bent knee wrap your arm around it. Then reach around your back and bend your torso towards your bent knee. Make sure your hand touch each other. Turn your head and look towards the ceiling and hold this pose for 20 sec to 1 min.

Pose: Virabhadrasana 2 Or Warrior 2 Pose

INSTRUCTIONS:

To perform this pose, stand in with one foot in front of you and the other foot behind you. Bend your knee bring your arms up and turn your head towards your bent knee. Make sure your arms are parallel to the ground.

Pose: Baddha Ardha Chandrasana Or Bound Half Moon Pose

INSTRUCTIONS:

To perform this pose stand on one foot and bend you other foot at the knee bring it back to your buttock. Hold on to your foot with the hand from the same side. Then bend your torso down on the side which your leg is flat on the floor. Stretch your arm out and touch the floor with your fingers and hold the pose. Switch sides and repeat the pose.

Pose: PARIVRTTA PARSVAKONASANA Or Twisted Side Angle

INSTRUCTIONS:

To perform this pose bend your right leg and put your left leg back. Then stretch your left arm to the ground and your right arm up to the ceiling. Remember to hold this pose with your finger tips on the floor. Switch sides and repeat the pose.

Pose: Utthita Eka Pada Rajakapotasana Or Standing One Legged King Pigeon Pose

INSTRUCTIONS:

To perform this pose stand on one leg and bend your other leg back so that your heel touches your butt. Hold this pose for 20 sec to 1 min. Use your arms to hold your leg and feel the stretch.

Conclusion

If you enjoyed this book, then I'd like to ask you for a favor;
Would you be kind enough to leave a review for this book on Amazon?
It'd be greatly appreciated!
Thank you and good luck!

Visit our website www.bestbuykindleebooksstore.com

Printed in Great Britain
by Amazon